S0-AFW-787
3 9082 09254 0056

DISCARD

SCIENCE FILES

earth

ISLANDS

Please visit our web site at: www.garethstevens.com
For a free color catalog describing Gareth Stevens Publishing's
list of high-quality books and multimedia programs,
call 1-800-542-2595 (USA) or 1-800-387-3178 (Canada).
Gareth Stevens Publishing's fax: (414) 332-3567.

Library of Congress Cataloging-in-Publication Data

Oxlade, Chris.
    Islands / by Chris Oxlade.
        p. cm. — (Science files. Earth)
    Includes bibliographical references and index.
    Contents: Earth's islands — What is an island? — Islands on the edge — Islands in
oceans — Reefs and atolls — Sun and storms — An island is born — The biggest island —
Island wildlife — Worlds apart — Island living — Using islands — Amazing facts.
    ISBN 0-8368-3568-9 (lib. bdg.)
    1. Islands—Juvenile literature. 2. Island ecology—Juvenile literature. [1. Islands.
2. Island ecology. 3. Ecology.] I. Title. II. Series.
    GB471.O95   2003
    551.42—dc21   2002030538

This North American edition first published in 2003 by
Gareth Stevens Publishing
A World Almanac Education Group Company
330 West Olive Street, Suite 100
Milwaukee, WI 53212 USA

Original edition © 2002 by David West Children's Books. First published in Great Britain
in 2002 by Heinemann Library, Halley Court, Jordan Hill, Oxford OX2 8EJ, a division of Reed
Educational and Professional Publishing Limited. This U.S. edition © 2003 by Gareth Stevens, Inc.
Additional end matter © 2003 by Gareth Stevens, Inc.

David West Editor: James Pickering
Picture Research: Carrie Haines
Gareth Stevens Editor: Alan Wachtel
Gareth Stevens Designer and Cover Design: Katherine A. Goedheer

Photo Credits:
Abbreviations: (t) top, (m) middle, (b) bottom, (l) left, (r) right

Ardea London Ltd.: Kurt Amsler (14-15b, 28-29); Jean-Paul Ferrero (29bl); François Gohier (24bl);
John Mason (22tr); McDougal Tiger Tops (13br); E. Mickleburgh (21br); D. Parer & E. Parer-Cook
(23r); Peter Steyn (9mr, 18-19b, 20-21b); Adrian Warren (22ml); M. Watson (18-19t, 29tl).
Corbis Images: Cover, 3, 4t, 5b, 8 (both), 8-9t, 10-11t, 12tr, 12-13, 16-17t, 22br, 25tl, 26r, 26-27,
27tr.
Ecoscene: Edward Bent (25bl); Sally Morgan (23bl).
Mary Evans Picture Library: 24bm.
Werner Forman Archive: 21tr.
Robert Harding Picture Library: C. Bowman (26bl); Martyn F. Chillmaid (25r); K. Gillham (28ml);
S. Grandadam (27tm); Robert Harding (16tr); Gavin Hellier (13tr, 19br); Gardar Igaliko (5tr); Louise
Murray (6tl); Nakamura (14-15t); Tony Waltham (11br, 20tr, 20-21m); 4b, 11ml, 14br, 17tr, 27br.
Papilio: Stephen Coyne (17br).
Popperfoto: 18bl, 29mr.

Printed in the United States of America

1 2 3 4 5 6 7 8 9 07 06 05 04 03

SCIENCE FILES

earth

ISLANDS

Chris Oxlade

Gareth Stevens Publishing
A WORLD ALMANAC EDUCATION GROUP COMPANY

CONTENTS

About half of the people on the huge island of New Guinea farm on the slopes of the island's high mountains.

Tropical islands are surrounded by spectacular coral reefs that provide homes for colorful fish and a wide variety of other sea creatures.

INTRODUCTION

From the high, jagged, jungle-covered islands that are located in the middle of oceans to the low, flat, grassy islands found close to the coasts of mainlands, the islands of the world are special places. Surrounding seas isolate them, creating worlds where unique plants and animals live, people's lifestyles remain unchanged for thousands of years, and natural resources are plentiful.

Greenland, the world's largest island, is mostly covered with ice. There are a few dozen settlements on its coast, where the snow melts in summer.

The Hawaiian Islands in the Pacific are formed by the summits of vast underwater volcanoes, some of which still spew out lava.

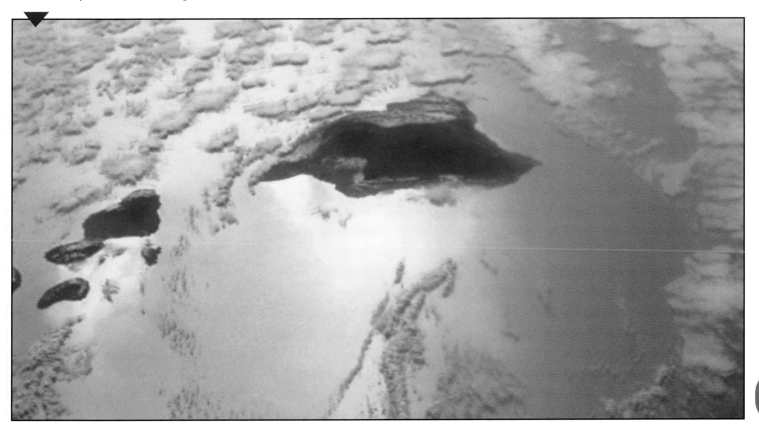

EARTH'S ISLANDS

The Pacific Ocean contains thousands of coral atolls like this one in the Solomon Islands.

There are many thousands of islands located around Earth. They range from rugged, mountainous islands in the icy polar regions to tiny coral atolls scattered throughout the warm tropical oceans near the equator.

ISLAND LOCATIONS

Most of Earth's large islands lie close to its continents. Islands lying further out to sea are usually smaller, although there are also plenty of small islands around coasts. Lakes, inland seas, and rivers also have islands in them.

CONTINENT OR ISLAND?

Australia and Antarctica are surrounded by sea, but they are not islands. Since they are parts of moving sections of Earth's crust, they are continents.

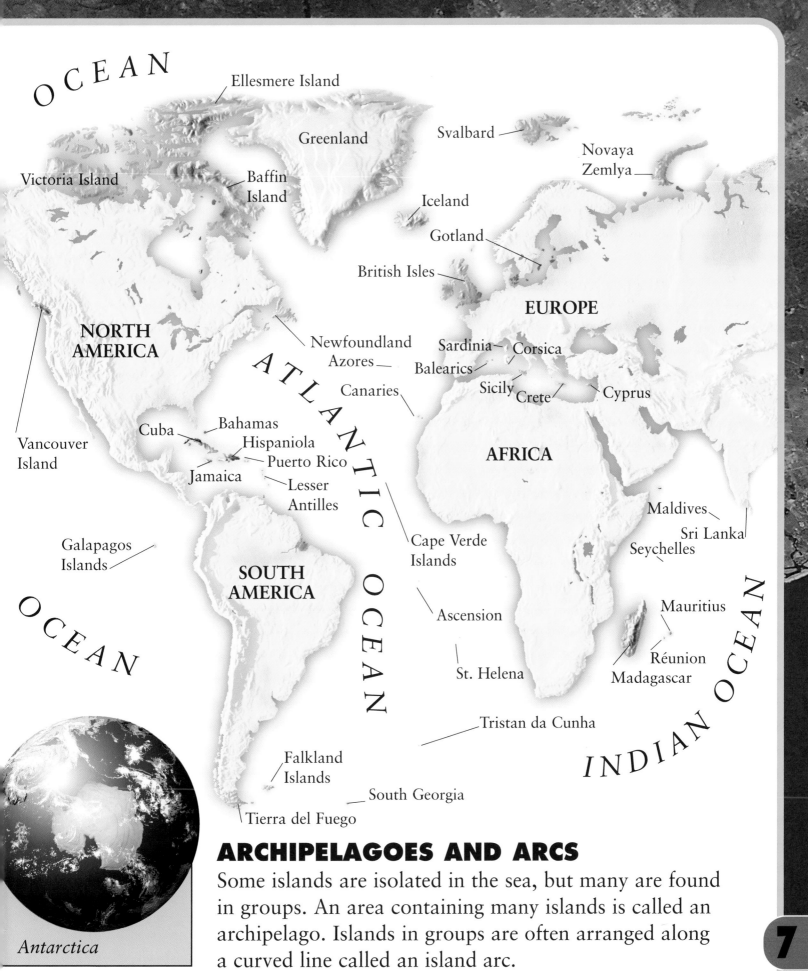

OCEAN

Ellesmere Island

Greenland

Svalbard

Novaya Zemlya

Victoria Island

Baffin Island

Iceland

Gotland

British Isles

EUROPE

NORTH AMERICA

Newfoundland

Azores

Sardinia — Corsica

Balearics

Canaries

Sicily — Crete — Cyprus

Vancouver Island

Cuba — Bahamas

Hispaniola

Puerto Rico

Jamaica

Lesser Antilles

AFRICA

ATLANTIC OCEAN

Maldives

Sri Lanka

Seychelles

Galapagos Islands

SOUTH AMERICA

Cape Verde Islands

OCEAN

Mauritius

Ascension

Réunion

Madagascar

St. Helena

INDIAN OCEAN

Tristan da Cunha

Falkland Islands

South Georgia

Tierra del Fuego

Antarctica

ARCHIPELAGOES AND ARCS

Some islands are isolated in the sea, but many are found in groups. An area containing many islands is called an archipelago. Islands in groups are often arranged along a curved line called an island arc.

WHAT IS AN ISLAND?

Very simply, an island is a piece of land that is completely surrounded by water. There is no official limit on how big an island can be, although Australia is considered to be a continent rather than an island.

Numerous small islands are part of a river delta in Madagascar that was formed from sediment washed down the river.

New Zealand is made up of two large oceanic islands formed by volcanic activity. Many of its volcanoes are still active.

ISLAND TYPES

There are two main types of islands — continental islands and oceanic islands. Continental islands are islands that are connected to a continent by a part of that continent that is underwater. Oceanic islands are islands out in the oceans that are not part of continents. The land that forms them rises up from the ocean floor.

MOST ISOLATED ISLANDS

The small group of volcanic islands that make up Tristan da Cunha are the most isolated islands on Earth. They lie in the South Atlantic, midway between South America and southern Africa. Only a few hundred people live on the biggest of these islands.

Tristan da Cunha

Andros Island is the largest island in the huge archipelago called the Bahamas.

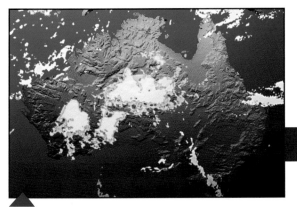

Australia and its surrounding islands make up the continent of Australasia.

MOVING CONTINENTS

The Earth's solid crust is made up of several huge parts, called tectonic plates, that float on the liquid rock underneath. Two hundred million years ago, there was one supercontinent, Pangaea. Gradual movement of the tectonic plates has created the continents we know today.

NORTH AMERICA
EUROPE —ASIA
AFRICA
SOUTH AMERICA
INDIA

PANGAEA

LAURASIA

GONDWANA

200 million years ago   180 million years ago   60 million years ago   Today

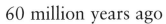

9

ISLANDS ON THE EDGE

Continental islands range in size from tiny offshore islands to enormous islands that make up substantial parts of continents, such as Greenland in North America and New Guinea in Australasia.

CONTINENTAL EDGES

At a continent's edge, Earth's surface drops steeply to the ocean floor, down a slope called a continental shelf. Shallow areas of sea before the shelf often contain continental islands.

Glaciers are remnants of the huge ice sheet that melted at the end of the last ice age.

FORMING CONTINENTAL ISLANDS

Continental islands are formed when the sea floods a section of coastal land, cutting off the high ground. The diagram on the right shows a section of hilly coast with a peninsula. When the sea floods the low-lying land, the hilltops on the peninsula are cut off from the mainland and create new islands off the coast, separated from the mainland by a shallow sea.

A peninsula extends out of the mainland.

MAINLAND AT TIME OF LOW SEA LEVEL

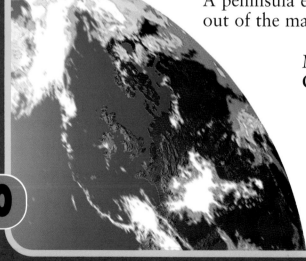

The British Isles are ▼ *part of Europe, but they are cut off from the mainland by the shallow North Sea.*

SEA LEVELS RISE

RISES AND FALLS

Continental islands are formed when coastal land is flooded. This happens either when the sea level rises or the land falls. Many of today's coastal islands were formed about 10,000 years ago, at the end of the last ice age, when a huge ice sheet melted and sea levels rose. Some coastal islands, such as the British Isles, were formed when sections of the land fell below sea level because of huge movements in Earth's crust.

The Orkney Islands, located to the north of Scotland, were created when sea levels rose 10,000 years ago.

When the sea floods a coast, peninsula hills become islands.

TIDAL ISLANDS

Some islands very close to the coast, such as St. Michael's Mount in Cornwall, England, are only islands temporarily. At low tide, they are linked to the mainland by exposed rocks and sand. At high tide, the link is submerged and they become islands again.

St. Michael's Mount, Cornwall, England

11

ISLANDS IN OCEANS

Isolated islands far out in Earth's oceans are called oceanic islands. Their land is not part of any continent but is formed by the summits of towering mountains that rise up from the ocean floor.

UNDERSEA VOLCANOES

Where magma leaks through cracks in the seabed, it builds underwater volcanoes. Erupting volcanoes eventually break the surface of the water and form new oceanic islands.

The island of Hawaii is the tip of a massive active volcano. Lava flows slowly to the sea, enlarging the island.

HOT SPOT ISLAND ARCS

Some oceanic islands form over hot spots. At a hot spot, magma breaks through the crust, building a volcano that may grow to form a volcanic island. As the tectonic plate over the hot spot slowly moves, the hot spot stays still, punching new holes in the crust and forming new volcanoes. Over millions of years, a line of hot spot volcanoes can create an island arc.

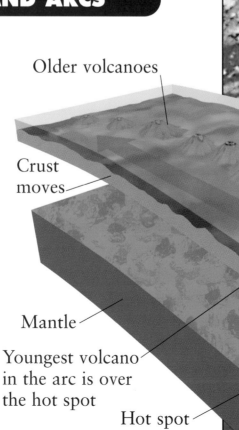

Older volcanoes

Crust moves

Mantle

Youngest volcano in the arc is over the hot spot

Hot spot

Iceland sits on top of the mid-Atlantic spreading ridge. The two halves of the island are slowly moving apart.

BOUNDARY ACTIVITY

Many oceanic islands form above the boundaries between tectonic plates. At a spreading ridge — a place where tectonic plates move slowly apart — magma moves up to fill the gap, often forming islands. Volcanic islands also form above boundaries called subduction zones, where one plate slides under another deep beneath the ocean.

KRAKATOA

In 1883, a devastating volcanic eruption blew away two-thirds of the island of Krakatoa, in Indonesia. The explosion caused a tidal wave, or tsunami, that drowned thousands of people on nearby islands.

Krakatoa is still an active volcano.

13

REEFS AND ATOLLS

The Pacific Ocean and the Indian Ocean are dotted with thousands of beautiful coral islands, such as the Maldives. The water around these islands teems with life.

A coral atoll with a dark-blue lagoon in the Caroline Islands, Micronesia, in the Pacific Ocean

CORAL REEFS AND ISLANDS

A coral reef is a ridge of coral that forms in shallow, tropical seas. Waves erode, or wear away, the coral, making sand that piles up on the reef to make an island. Tropical plants take root in the sand, helping to bind it together.

The Pacific island of Bora Bora is made up of a central volcanic peak surrounded by numerous coral reefs and islands.

CORAL POLYPS

Coral is made from the skeletons of millions of tiny animals called polyps. Different polyps form colorful corals in the shape of branches, fans, and ridges.

A coral reef

ATOLLS AND LAGOONS

Coral reefs can form offshore, where they are called barrier reefs, or close to shore, where they are called fringing reefs. An atoll is a ring of coral reefs and islands surrounding a central lake, called a lagoon. Because of the way the reefs grow, the land of a coral island is never more than a few yards above sea level.

CORAL ATOLL FORMATION

1) A coral atoll begins to form when an underwater volcano begins to grow over a hot spot on the ocean floor. 2) The volcano continues to erupt until it breaks the surface, creating a new island.

3) A fringing coral reef grows in the shallow water around the island, and the volcano becomes extinct. 4) The cone of the volcano begins to erode away, but the coral continues to grow. 5) Eventually the volcano cone disappears completely, leaving behind a coral atoll with a central lagoon.

SUN AND STORMS

Earth's islands experience the whole range of the planet's climates, or patterns of weather, from polar to tropical. Some islands are so large that the climate in the center is quite different from the climate on the coasts.

CLOUDS AND RAIN

The seas that surround islands have a huge effect on island climates. In hot climates, it can rain almost every day on an island, as heat from the land encourages clouds to form in the moist air over the ocean.

Hurricane winds, which can reach speeds of 155 miles per hour (250 kilometers per hour), blow palm trees on the island of Bermuda.

OCEAN CURRENTS

Island climates are often affected by ocean currents, which bring warm or cold water to different parts of the globe. The Gulf Stream, for example, carries warm water from the Caribbean to islands in the North Atlantic, such as the British Isles. This makes these islands' climates milder than other islands so far to the north.

GULF STREAM

NIGHT AND DAY

The Sun warms the land, and the land, in turn, warms the air above it. This air rises, creating clouds and onshore breezes.

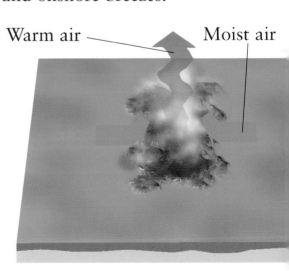

Warm air — Moist air

DAY

ISLAND STORMS

Islands out in the oceans are unprotected from storms that bring destructive winds and flooding. The islands in the Caribbean are in the path of hurricanes sweeping in from the Atlantic, and islands in the Philippines are in the path of Pacific typhoons.

The tight swirl of a hurricane system over the Atlantic Ocean, photographed from space. The hole in the middle is called the eye.

Islanders in Indonesia take cover from a seasonal storm.

After sunset the land cools, taking heat from the air above. This air sinks, creating offshore breezes and making the clouds scatter.

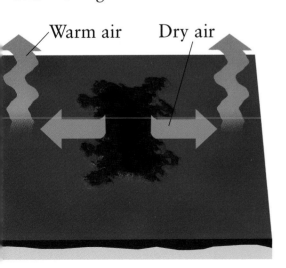

Warm air   Dry air

NIGHT

Clouds forming over St. Kitts in the Caribbean

AN ISLAND IS BORN

Every so often, a new island appears. The most spectacular births are those of volcanic islands, which burst unexpectedly out of the oceans. New islands are also formed by growing corals, as well as by the action of rivers and tides.

BIRTH OF SURTSEY

In 1963, smoke and steam suddenly started to rise from the sea off the south coast of Iceland. The island of Surtsey was being born. Just four days later it was about 1/3 mile (1/2 kilometer) across. Surtsey stopped erupting in 1967, and the new island was soon colonized by plants, birds, and insects.

On the islands in the delta of the River Betsiboka, in Madagascar, the land is fertile, but farming can be hazardous because of floods.

Surtsey as it is today (background), ▶ *and as it was when it first emerged from the ocean in 1963 (below)*

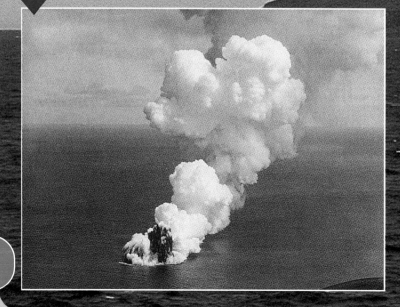

DELTA ISLANDS

A delta is a fan-shaped piece of land that is cut through by channels. It forms where a river dumps sediment at its mouth. When a river floods, the channels of a delta often change position, destroying some islands and creating others.

VOLCANIC BIRTH

Volcanic islands, such as Surtsey, form when magma from beneath Earth's crust breaks out, forming a volcano. The eruption gradually builds up a cone-shaped mountain of lava. When the volcano breaks the surface, layers of ash are deposited, along with lava.

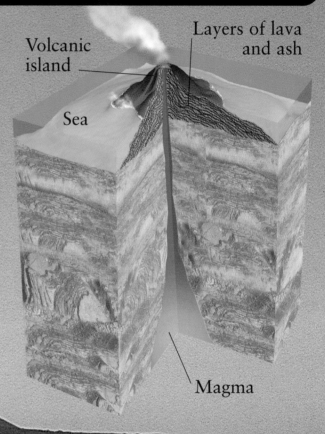

Volcanic island

Layers of lava and ash

Sea

Magma

DEATH OF AN ISLAND

Atlantis is a legendary island that is said to have sunk into the sea thousands of years ago. Some people believe that Atlantis actually existed and that it was a volcanic island destroyed by catastrophic eruptions. One possible site for Atlantis is the volcanic island of Santorini, which is the remains of a larger island.

Santorini in the Greek islands

THE BIGGEST ISLAND

Lying between the North Atlantic Ocean and the Arctic Ocean is a vast island, permanently covered in a thick sheet of ice. This is Greenland, the largest island in the world. Despite the fact that Greenland sits almost completely inside the Arctic Circle and has a very cold climate, about 50,000 people live around its coast.

The ice on Greenland's ice sheet averages .9 mile (1.5 km) thick, but it reaches twice this thickness in the center of the island.

GREENLAND

Greenland, which is part of North America, measures 1,650 miles (2,655 km) from north to south and 800 miles (1,290 km) from east to west. It is almost two-thirds the size of Australia.

Ellesmere Island

Baffin Bay

Baffin Island

GREENLAND

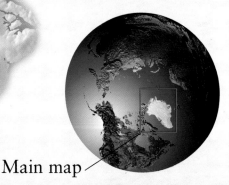

Main map

THE ICE SHEET

A thick sheet of ice covers 85 percent of Greenland. The ice sheet never melts, and the summits of Greenland's mountains poke up through it. There are numerous islands and fjords around Greenland's coasts. Glaciers flow down from the ice sheet into the fjords, filling the fjords with icebergs. Most of the coast, however, is free of ice and snow.

VIKING SETTLEMENTS

Greenland was named by the Vikings, who first settled there in about A.D. 985, led by Eric the Red. The Viking colony grew to about 3,000 people before the climate cooled in the 1400s and made it impossible for them to stay.

The remains of Brattahild, Greenland

PEOPLE OF GREENLAND

Most Greenlanders live in towns on the island's southwestern coast. They are descended from Inuit and European settlers. A few people live traditionally by hunting, but most people on Greenland work in the island's fishing industry.

All of Greenland's settlements are on its coast.

An Inuit man wearing sealskin clothes

21

ISLAND WILDLIFE

The nature of an island is that it is separated from other areas of land. This isolation has made it possible for some unique animals and plants to evolve on islands.

Plant seeds often wash up on island shores.

1) Wind carries seeds and insects.

2) Small animals and plants are carried by ocean currents on floating debris.

The Komodo dragon is a huge lizard.

Orangutans are under threat of extinction because their rain forest habitats on Borneo and Sumatra are being destroyed.

ISLAND EVOLUTION

When a piece of land is cut off to create an island, species on it often evolve differently from animals on the mainland. The Komodo dragon, for example, evolved only on some islands in Indonesia.

COLONIZING AN ISLAND

When new volcanic islands or coral atolls are formed, the land is lifeless. But plants and animals soon begin to colonize them. Birds and insects fly there, and plant seeds are carried by the waves or wind. Large land animals cannot colonize oceanic islands.

3) Bird droppings deliver seeds.

4) Humans bring animals and plants that may destroy an island's delicate balance.

GIANT PLANTS

The coco-de-mer, a palm tree from the Seychelles, has the largest seeds of any plant. These seeds drift across the sea and take root where they land.

A sprouting coco-de-mer

FLIGHTLESS BIRDS

Many species of birds only fly to escape predators. On remote islands where there are no predators, flightless birds, such as the kiwi of New Zealand, have evolved. Unfortunately, many flightless birds, such as the kakapo, also of New Zealand, are hunted by domestic animals introduced into the island environment by humans.

Frigate birds fly to Aldabra, in the Indian Ocean, to breed. The birds' eggs and young are safe there because of the lack of predators.

23

WORLDS APART

The Galapagos Islands, in the Pacific Ocean, and Madagascar, in the Indian Ocean, are treasure chests of nature. On each of these islands, there are strange species of animals that are not found anywhere else in the world.

THE GALAPAGOS

Among the amazing kinds of animals that have evolved on the Galapagos Islands are the marine iguana, the only species of lizard that lives in the sea, and the giant tortoise. Fourteen species of finch also live on the islands, each slightly different from the others.

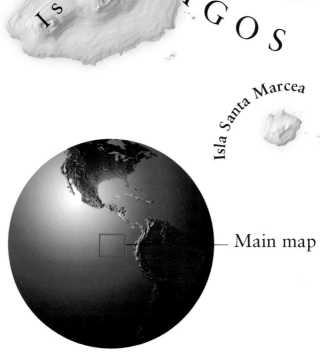

THE GALAPAGOS

Isla Pinta

Isla Marchena

Isla San Salvador

Isla Fernanda

Isla Isabela

Isla Baltra

Isla Pinzon

Isla Eden

Isla Santa Marcea

Main map

Charles Darwin, the British naturalist, visited the Galapagos Islands in 1835. Studying the island's unique animals, including its finches, helped him to develop his theory that animals evolve by natural selection.

24

The marine iguana feeds on algae in the sea off the Galapagos Islands. No other lizard does this.

Isla Santa Fé

Isla San Christobal

ISLANDS

Isla Española

Madagascar is the fourth largest island in the world. It split away from the rest of Africa about 100 million years ago.

MADAGASCAR

Most of the animals and plants in Madagascar are unique. The island's most famous animals are its lemurs. All the species of lemur are now threatened because their forest habitats have been destroyed.

ISLAND VISITORS

Wildlife enthusiasts visit the Galapagos and Madagascar to see the islands' unique animals for themselves. The money they spend helps to set up conservation programs for endangered species.

Tourists encounter giant tortoises in the Galapagos.

Lemurs, such as this ring-tailed lemur, use their prominent tails for signaling to each other.

25

ISLAND LIVING

For thousands of years, people have moved to the world's islands in search of new land for farming and settlement or simply to explore new ways of life.

ISLAND LIFESTYLES

On remote islands — such as those in Polynesia, in the South Pacific — people live traditional lifestyles. They survive by fishing, raising animals, and growing crops for food. The Polynesians colonized these islands tens of thousands of years ago, traveling vast distances across the Pacific Ocean in canoes. Isolated island life is not always easy, and many young people leave the Polynesian islands to find work on the mainland.

On islands such as Hong Kong and Singapore, people live modern lifestyles in bustling cities. Hong Kong has become a major center for international commerce.

Polynesians are excellent sailors. They fish and travel in narrow dugout canoes.

Borneo is the world's third largest island. Most of its inhabitants are people known as Dayaks. Many still practice traditional crafts, including weaving beautiful cotton cloth.

CONNECTING ISLANDS

For island countries, such as Japan, that have developed into commercial and industrial centers, fast transportation links are vital. Road and rail bridges and tunnels have been built to link islands to each other and to the mainland.

Kyoto Multiple Bridge, Japan

Most of the six million people of New Guinea live in villages. In lowland villages, thatched houses are built on stilts to keep them dry in the rainy climate.

USING ISLANDS

The islands of the world boast a wealth of resources, from abundant sea life and rich farming land to rare minerals and geothermal energy. Beautiful island scenery and beaches draw tourists by the millions.

The fertile, black volcanic soil and the mild Atlantic climate of the Canary Islands make the islands ideal for growing grapes that are made into wine.

ISLAND PRODUCTS

The main industries on islands are mariculture, such as fish farming and pearl farming, and fishing. Islanders fish for food and also export their catches. Some island climates are good for growing crops and raising animals. The rocks of volcanic islands often contain useful minerals, such as gold and gemstones, that are mined and exported to the mainland or to other countries.

Leatherback turtles lay their eggs on tropical beaches. Unfortunately, their nesting sites are being badly disrupted by tourists.

Coral is collected from reefs for building material and to make souvenirs for tourists. Coral reef harvesting destroys the precious reef habitats that have taken thousands of years to grow.

ATOMIC ATOLLS

The remoteness of the Pacific islands is not always in their favor. The people of Bikini Atoll were moved to other islands so that the United States could test nuclear weapons there. The U.S. is now paying for the islanders to move back to their native areas.

Nuclear bomb test at Bikini Atoll

Beachcomber Island is a beautiful vacation spot. It is one of the hundreds of islands in Fiji with luxury lodgings.

ISLAND TOURISM

Island landscapes — from the white sandy beaches of the Caribbean to the rugged volcanic features of Iceland — make islands popular tourist destinations. The economies of many islands, such as the Balearics, located off Spain, depend on tourism. The thousands of tourists in new hotels, however, also create problems, such as water shortages and pollution.

AMAZING FACTS

THE LARGEST ISLAND	Greenland is the largest island in the world. It has a total area of about 840,000 square miles (2,175,600 square kilometers). More than 695,000 square miles (1,800,000 sq km) of Greenland are buried under ice. The island's coastline is about 3,600 miles (5,800 km) long.
THE BEST OF THE REST	The remaining islands on Earth's top five (by area) list are as follows: New Guinea (312,100 square miles/808,510 sq km), Borneo (292,220 square miles/757,050 sq km), Madagascar (229,350 square miles/ 594,180 sq km), Baffin Island (196,090 square miles/508,000 sq km).
THE LARGEST BY CONTINENT	The largest islands on each continent, except Antarctica, are as follows: North America: Greenland (*see above*); South America: Tierra del Fuego, 18,140 square miles (47,000 sq km); Europe: Great Britain, 88,730 square miles (229,870 sq km); Africa: Madagascar (*see above*); Asia: Borneo (*see above*); Australasia: New Guinea (*see above*).
TALLEST ISLAND	The tallest island, measured from the ocean floor, is the main island of Hawaii. Its highest volcano, now extinct, is Mauna Kea. Its summit is 13,797 feet (4,205 m) above sea level, and an astonishing 30,185 feet (9,200 m) above the ocean floor. That makes Hawaii taller than Mount Everest. Kilauea, also on Hawaii, is the world's most active volcano.
THE COUNTRY WITH MOST ISLANDS	Indonesia is made up of about 13,600 islands. Many of these are less than two-thirds of a mile across, but Indonesia also includes parts of the huge islands of New Guinea and Borneo, making up a total area about the size of Greenland. About 6,000 of the islands are inhabited.
THE MOST ISOLATED INHABITED ISLAND	Tristan da Cunha, with a population of just 300 people, lies in the South Atlantic, about 1,616 miles (2,600 km) from Cape Town, South Africa, and about 1,864 miles (3,000 km) from Rio de Janeiro, Brazil, making it the most isolated inhabited island in the world.
THE MOST AT RISK	The average level of land in the Maldives, in the Indian Ocean, is only about 8 feet (2.5 m) above sea level. If global warming causes the sea to rise, the Maldives will be the first islands to disappear under the waves.

GLOSSARY

archipelago: an area containing many islands; a large group of islands.

fjord: a long, narrow, steep-sided valley that runs inland from the sea and is filled with seawater.

global warming: the gradual increase in the average temperature of Earth's atmosphere.

ice age: a period when Earth's temperature dropped and thick sheets of ice and glaciers extended south, covering much of North America, Europe, and Asia.

lava: the name given to magma when it leaves a volcano.

magma: hot, molten rock that lies just underneath Earth's solid crust.

peninsula: a long strip of land, almost entirely surrounded by water, but connected to the mainland at one end.

Polynesia: a huge group of islands in the Pacific, stretching from New Zealand to Easter Island.

spreading ridge: a boundary between two tectonic plates at which the plates gradually move apart.

subduction zone: a boundary between two tectonic plates at which one plate slides underneath the other.

tectonic plate: a huge section of Earth's crust that floats on and moves over Earth's mantle. Earth has several tectonic plates.

MORE BOOKS TO READ

Coral Reef. Webs of Life series. Paul Fleischer (Benchmark Books)

Galapagos: Islands of Change. Lynne Borne Meyers (Hyperion Press)

Islands. Mapping Earthforms series. Catherine Chambers (Heinemann Library)

People of the Islands. Wild World series. Colm Regan (Raintree Steck-Vaughn)

WEB SITES

Explore Madagascar
www.pbs.org/wgbh/nova/madagascar/explore

Surtsey
www.watson1999-69.freeserve.co.uk/surtsey

Due to the dynamic nature of the Internet, some web sites stay current longer than others. To find additional web sites, use a reliable search engine with one or more of the following keywords: *archipelagoes, Galapagos, Greenland, Hawaii, hot spots, Krakatoa, Madagascar, Polynesia.*

INDEX